A PRACTICAL GUIDE TO MASTERING NUMERACY TESTS

GRADES 4 – 6

Volume 2: MATHEMATICAL OPERATIONS AND RELATIONS

Jasmin C. Alexander

ISBN-10: 153027415X

ISBN- 13: 978-1530274154

Printed in the United States of America

1

TABLE OF CONTENTS

OPERATIONS AND RELATIONS

1. ADDITION

2. SUBTRACTION

3. MULTIPLICATION

OPERATIONS AND RELATIONS

– Some Things to Remember

MASTERY TEST 1

MASTERY TEST 2

MASTERY TEST 3

CERTIFICATE OF ACHIEVEMENT

INDEX

NOTES

This book belongs to

_____

*School:*_____

Class: _____

❖ *Practice makes perfect…*

❖ *Your attitude determines your altitude…*

MATHEMATICAL OPERATION AND RELATIONS

Chapter 1 ADDITION

The four (4) mathematical operations are Addition, Subtraction, Multiplication and Division.

1.1 Understanding Addition:

The addition sign is shown as '+' in number sentences. When something (i.e. a number) is added to another number, the value is increased. However, when nothing (i.e. zero) is added to another number, the value remains the same. The Properties of Addition, discussed below, will explain this in more detail.

Some words that mean to add are: *sum, plus, add, increase, total, got more, altogether* and *in all*.

1.2 Properties of Addition:

> ### Identity Property:

The identity property shows that if you add zero (0) to any number, it will remain the same. For example:

a) $45 + 0 = 45$ ii) $7890 + 0 = 7890$ iii) $0 + 12 = 12$

➢ Commutative Property:

The commutative property of addition shows that it does not matter which addend is added first; what is important is that the digits in the numeral are placed in their correct place value column before adding. Therefore:

i) 2 + 7 is the same as 7 + 2

$$\begin{array}{r} 2 \\ +7 \\ \hline 9 \end{array} \qquad \begin{array}{r} 7 \\ +2 \\ \hline 9 \end{array}$$

ii) 568 + 24 +1 = 24 + 1 + 568

$$\begin{array}{r} \overset{1}{568} \\ 24 \quad \cancel{13} \\ + \quad 1 \\ \hline 593 \end{array} \qquad \begin{array}{r} \overset{1}{24} \\ 1 \quad \cancel{13} \\ + 568 \\ \hline 593 \end{array}$$

➢ Associative Property

The associative property shows that the sum (the answer) remains the same regardless of how numbers are grouped to be added. Example:

i) (1 +5) + 8 = 1 + (5 + 8)

$= \quad 6 + 8 = 1 + 13$

$= \quad 14 \quad = 14$

ii) (7+ 5) + (6+3) = (7+6) + (5+3)

$= \quad 12 \quad + \quad 9 \quad = \quad 13 + \quad 8$

$= \quad 21 \qquad \qquad = 21$

As you can see, both sides of each problem were equal even though the numbers were grouped differently.

    i. $(5+6) + 2 = 5 + (6 + 2)$ _____

    ii. $19 + 0 = 19$ _____

    iii. $6 + 2 = 2 + 6$ _____

1.3 Parts of the Addition Number Sentence

The main parts of an addition problem are the **addends** and the **sum.** The addends are the numbers that you are adding and the **sum** is result (answer) that you get when these numbers are added together. It is the **largest number** in an addition problem. It usually comes at the end of the problem after the equal sign. Example

$$12 + 45 = 57$$

                  addends   sum

1.4 Finding the Missing Part of an Addition Number Sentence:

To solve any given addition problem, you will have to find either the sum or an addend

> ➢ To find the sum, **add the addends** together.
>
> (e.g. $6 + 8 = \boxed{14}$)
>
> ➢ To find a missing addend, **subtract the given addend(s)** from the sum. For example:

$$\boxed{?} + 5 = 9$$
$$\boxed{?} = 9 - 5 = 4$$

Answer these questions below correctly

1) In the number sentence: **14 + 10 + 43 = 67,** the number 43 is a/an _____ and the number 67 is the _____.
(sum, addend)

2) $5 + 8 + 4 = \boxed{}$ 3) $\boxed{} + 16 = 59$

4) $5 + \boxed{} = 37$ 5) $47 + \boxed{} = 47$

9

1.5 Adding Numbers without Regrouping:

When adding numbers, you must:

> ➢ Write the numbers down vertically if they are too big to add horizontally.

> ➢ Make sure that each number's digit is in its correct place value column

> ➢ Then add the numbers, starting from the columns on the right and move leftward to sequential columns. (If the sum of the numbers is less than ten, then there is no need for regrouping).

> ➢ Add all the digits in the Ones' column then put their sum in the Ones' column answer space below of the numbers added.

> ➢ Repeat the same steps for number in the Tens' column and so forth. For example:

Question: What is the sum of 423, 13 and 8102?

```
Th. H. T. Ones
      4 2 3
        1 3
  + 8 1 0 2
  ─────────
    8 5 3 8
```

     i.   453 + 1203 +1032

     ii.   Add: 241, 100 and 46

     iii.   36201+ 21577

     iv.   1102 + 263 + 5612

1.6 Adding Numbers with Regrouping

As stated before, the system that we use is the 'Base- 10 System.' Hence, ten groups of Ones equal one Ten; ten groups of Tens equal One Hundred; ten groups of a hundred equal One

Thousand etc. Therefore, the largest number digit that is placed in any place value column is nine.

- When adding numbers, follow the same steps as stated above

- However, if the sum of number added in any column is more than nine (9), you have to put it outside and regroup it. For example:

Question 1: $1037 + 6784 =$

$$
\begin{array}{r}
\overset{1}{}\\
1\,0\,3\,7 \quad \text{11}\\
+\ 6\,7\,8\,4 \quad \text{12}\\
\hline
7\,7\,2\,1
\end{array}
$$

Question 2: What is the sum of 467, 13 and 8709?

Th.	**H.**	**T.**	**Ones**
1	1	1	
4	6	7	
	1	3	11
+ 8	7	0	9 19
9	1	8	9

or

Th.	**H.**	**T.**	**Ones**
1	1	1	
8	7	0	9
	4	6	7 19
+		1	3 11
9	1	8	9

Remember:

➢ *It does not matter which number is added first. What is important is that each number's digit is in its correct place value column.*

> ➤ *If the numbers of a column add up to be more than '9',*
> *put the answer outside and regroup it.*

Solve these number sentences correctly:

1) $754 + 677 + 1230 =$

2) $9899 + 1002 + 5 + 12 =$

3) $21368 + 4597255 + 1000 =$

4) $10 + 1000 + 100 + 39 + 8984 =$

1.7 Solving Worded Problems

When solving worded problems, it is good to read the question carefully and look for key words that will give you a hint on what Mathematical Operation you need to carry out in order to solve the problem.

Some Key Words That Mean To Add

- ❖ *plus*
- ❖ *add*
- ❖ *increase*
- ❖ *total*

- ❖ sum
- ❖ *got more*
- ❖ *altogether*
- ❖ *in all*

Exercise 5

Read the worded problems below. Underline the key word(s) that tell(s) you what to do. Then solve correctly.

1. What is the sum of 476, 852 and 999?

2. Find the total of the numbers between 1 and 10

3. Ronan had 600 oranges. He got 368 more oranges. How many oranges does he have altogether?
4. Increase 26 by709

5. What is 534 plus 843, plus 78 equal?

6. David had 43 apples. He got 588 more from Sally and 85 more from Pamela. How many apples does he have altogether?

1.8 Adding Numbers with Decimal Fraction:

When adding numbers with a decimal fraction, remember:

➢ The decimal point separates the whole number from the fraction/part of a whole.

➢ All the digits should be placed in their correct place value column.

➢ All the decimal points are placed directly under each other.

➢ If some numbers have two digits after the decimal point and others do not, you can put a '0' in their tenths or hundredths column to take up the missing digit space. This may make adding the numbers easier.

➢ The rules remain the same as adding whole numbers columns; if the sum of digits in a column is less than ten (i.e. 9 or less) write it in the same answer column

➢ If the sum of digits in a column is more than nine (i.e. 10 or more) write it outside and regroup it. For example:

Question: Find the 6.9, 34. 33, 107.2 and 615

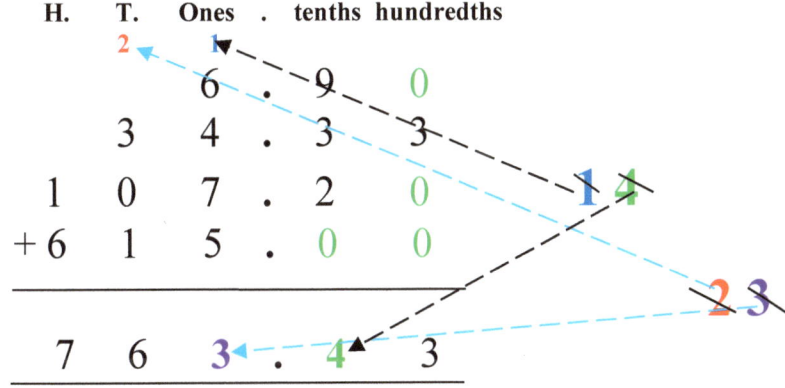

Note:

➢ The arrows show where each digit goes when it is regrouped.
➢ When we added the digits in the **tenths** column, we got 14. This number had to be regrouped because it was more than 9.

➤ **14 tenths** = **1 One** + **4 tenths,** so we marked off the **4** and placed it in the tenths column, then we marked off the **1** and placed it in the Ones column.

➤ When we added the digits in the **Ones** column, we got **23.** This number had to be regrouped also because it is more than 9.

➤ **23** Ones = **2 Tens** + **3 Ones**, so we marked off the **3** and placed it in the **Ones** column, then we marked off the **2** and placed it in the **Tens** column.

Exercise 6

Solve these problems below correctly

1) 22.3 + 42. 1 + 10.2 =

2) Add 943.32 4, 2.114 and 67.900

3) Add 34.78, 240.5 and 4.6

Chapter 2 SUBTRACTION

2.1 Understanding Subtraction

Subtraction is the opposite of addition. The subtraction sign is shown as a 'dash-like symbol' (–) in a number sentence. To subtract means to 'take away'. If you 'take away' something from what you have, then what you have remaining would be less than what you had at first. However, if you took nothing away (i.e. 0), then what you have will remain the same.

The properties of subtraction will explain subtraction in more detail. For simplicity, we will refer to the number that

2.2 Properties of Subtraction

➢ When zero (0) is subtracted from any number, the number remains the same. This is known as the Identity Property. For example: $9 - 0 = 9$, $543 - 0 = 543$.

➢ If you take away a number that has the same value as the number that you are taking away from, the remainder will be zero. For example $25 - 25 = 0$, $3,789 - 3,789 = 0$

➢ Unlike addition, subtraction of whole numbers is not commutative. For example, $8 - 3 = 5$; but $3 - 5$ is not possible.

➤ Unlike addition, subtraction of whole numbers is not associative.

$7 - (4 - 2)$ is not the same as $(7 - 4) - 2$.
$7 - (4 - 2) = 7 - 2 = 5$ $(7 - 4) - 2 = 3 - 2 = 1$

2.3 Parts of the Subtraction Number Sentence

The three main parts of a subtraction problem are the *minuend*, *subtrahend* and the *difference*.

➤ The **minuend** is the number from which you are subtracting. For simplicity, we can refer to it as the '**Have**' number (the number showing how much you/they have) according to the number sentence).

➤ The **subtrahend** is the number that is being subtracted from the minuend. It is the number being taken away from what 'you' have.

➤ The **difference** is the amount that remains after subtraction.

For example:

$648 - 123 = 525$ $3995 - 451 = 3544$

H. T. Ones		
6 4 8	⟵	minuend
− 1 2 3	⟵	subtrahend
5 2 5	⟵	difference

Th. H. T. Ones		
3 9 9 5	⟵	minuend
− 4 5 1	⟵	subtrahend
3 5 4 4	⟵	difference

2.4 Finding Any Missing Part of a Subtraction Number Sentence

As shown above, there are three main parts in a subtraction problem. The minuend is positioned at the beginning (first) in a subtraction number sentence and is usually the biggest number. The subtrahend is placed in the middle after the subtraction sign
and the difference is shown after the equal sign as it shows what remains after subtraction took place.

> To find the ***difference,*** take away the *subtrahend* from the minuend (what 'you' have). For example

$$7 - 5 = \boxed{2}$$

minuend   subtrahend   difference

> To find the ***subtrahend,*** take away the *difference* from the minuend. For example, using the same problem above

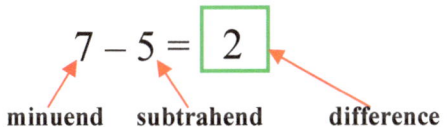

minuend   subtrahend   difference

> To find the ***minuend***, add back the number that you subtracted (i.e. *subtrahend)* to the amount that you have remaining (i.e. *difference*). For example, using the same problem above

$$\boxed{} - 5 = 2 \longrightarrow \boxed{} = 2 + 5 \longrightarrow \boxed{} = 7$$

minuend   subtrahend   difference

➤ For bigger numbers, these problems can be solved by writing the number sentence vertically. For example:

$$765 - \boxed{} = 251, \text{ what is } \boxed{} ?$$

From this problem, we are asked to find the **subtrahend.** The rule for finding the subtrahend (i.e. How much was taken away) is to take away the difference from the minuend. Therefore,

```
    H. T. Ones
      7 6 5
  -   2 5 1
      5 1 4  ←——— subtrahend
```

<mark>**Exercise 7**</mark>

<mark>**Find the missing part for each of the subtraction problem below**</mark>

i. $14 - 8 = \boxed{}$ ii. $9 - \boxed{} = 5$

iii $\boxed{} - 46 = 32$ iv. $9648 - 413 = \boxed{}$

2.5 Subtracting Numbers without Regrouping

If the value of each digit in the minuend is greater than the value of each digit in the subtrahend, then the subtrahend (number to be taken away) can be deducted from the minuend (number 'you' have) without regrouping. For example

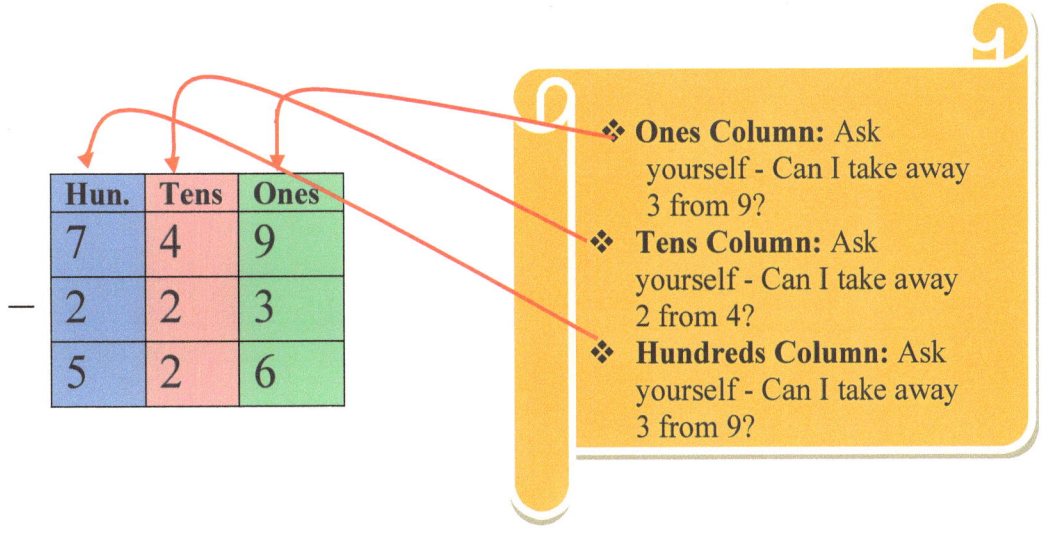

Hun.	Tens	Ones
7	4	9
2	2	3
5	2	6

❖ **Ones Column:** Ask yourself - Can I take away 3 from 9?

❖ **Tens Column:** Ask yourself - Can I take away 2 from 4?

❖ **Hundreds Column:** Ask yourself - Can I take away 3 from 9?

Solve Correctly

H.	T.	Ones
6	7	5
1	3	5

H.	T.	Ones
8	4	9
6	1	8

H	T	Ones
7	7	5
4	7	0

Remember: *Always begin to solve any mathematical problem by starting with the **Ones Column**.*

2.6 Subtracting Numbers with Regrouping

In the subtracting of whole numbers, the *minuend* should be bigger than or equal to the *subtrahend*. To put it simply – 'What you have should be bigger or equal to what you are going to take away, if the *difference* is to be a whole number.

➢ It further goes to say that you cannot take away a number or an amount that is bigger than what you have.

➢ In any column (Ones, Tens, Hundreds etc.), if the digit of the minuend (Have Number) is smaller than the digit of the subtrahend, then you will have to take one (1) from the digit of the next minuend that is in the column with a greater place value – This is known as regrouping.

Tens	Ones
2	
~~3~~	^{1}6
−	7
2	9

❖ **Ones Column**: You cannot take away 7 from 6.
❖ Therefore, you go to the **Tens Column** and take **1** ten from the '**3**' tens.
❖ Put the '**1**' ten you took besides the '6' ones (10 + 6 = ^{1}6)
❖ '**2**' tens remain, so mark off the '3' and put '2' above to show what remained

➢ If the digit of the next minuend (e.g. Tens Column digit) is zero (0), and you need to regroup, then you will have to take one (1) from the next column with a greater place value (i.e. Hundreds Column).

➢ However, you cannot skip over the digit of the minuend with zero (0) on your way to take the one (1) where it is needed (You cannot skip over the Tens column to take the '1' from the Hundreds column to the ones column)

➢ You will need to put the one (1) from the Hundreds column next to the zero(0) in the Tens column so it becomes a ten (10), then take one (1) from that ten and carry it to the Ones column. The digit in the Tens column will now be nine (9). See the example below.

```
 H.  T. Ones
  6   9
  7  ¹0  ¹4
-  5  2   8
-----------
     1  7  6
-----------
```

Solve :

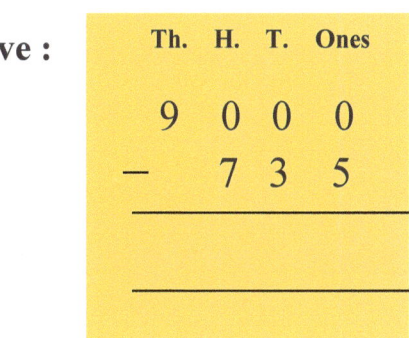

```
 Th.  H.  T. Ones
  9   0   0   0
-         7   3   5
-------------------


-------------------
```

2.7 Solving Worded Problems

When solving worded problems, there are certain key words that will give you hint on what mathematical operation is required. Some key words that mean to subtract are listed below.

Words That Mean To Subtract

❖ minus	❖ have remaining
❖ take away	❖ decrease by
❖ give away	❖ reduce by
❖ difference between	❖ less
❖ have left	❖ subtract

In some worded problems you may see words that implies that something was taken from the 'have number' or the minuend. In such cases, you should subtract. Some examples of these words are ... **had ... lost; had ...ate; had... sold** etc.

As you can see, if you had something and you lost, sold or ate some, it will become *less*, therefore these words usually imply to subtract.

Read the worded problems below. Underline the key word(s) that tell(s) you what to do. Then solve correctly in your workbook.

1. What is 248 take away 33?

2. Sally had 821 apples. She sold 305 apples. How many apples does she have remaining?

3. What is the difference between 499 and 192?

4. Reduce 2000 by 564

5. If 2503 was reduced by 1373, how much will remains?

6. Harry had 45390 beads. He lost 798 beads. How many beads does he have left?

Chapter 3 MULTIPLICATION

3.1 Understanding Multiplication

Multiplication is repeated addition of the same number. The multiplication sign '**x**' is written between two numbers in a number sentence. It is sometimes read as 'times'. If you learn your multiplication tables, you will be able to solve multiplication problems quickly. When you multiply a number, it value is increased by multiples of that number or by repeated addition of that number by the amount of times (x) specified. However, any number that is multiplied by '0' equals zero.

3.2 Properties of Multiplication

➢ Identity Property

The identity property shows that any number multiply by '1' equals itself.
For example, 4 x 1 = 4;   73 x 1 = 73,   96 x 1 = 96

➢ Commutative Property

In multiplication, the order in which numbers are multiply does not matter because it does not change the product or result. For example, five times seven equal thirty-five; likewise, seven times five equal thirty-five.
 (i.e. 5 x 7 = 7 x 5)

➤ **Associative Property**

The associative property shows that the order in which numbers are grouped for multiplication is not important because it will

not change the value of the numbers to be multiplied. For example $3 \times (2 \times 4) = (3 \times 2) \times 4 = 2 \times (4 \times 3)$

$= 3 \times 8 = 24 = 6 \times 4 = 24 = 2 \times 12 = 24$

➤ **Multiplying by '0'**

Any number multiply by '0' equal '0'

e.g. $2 \times 4 \times 0 = 0$ $5007 \times 0 = 0$

➤ **Distributive Property:** Multiplying a number by the sum of two or more numbers is the same as multiplying each addend by the number then adding their products. For example

$2(6 + 3) = \quad \longrightarrow \quad (2 \times 6) + (2 \times 3)$

$= 2 \times 9 \quad = 18 \quad \longrightarrow \quad 12 + 6 = 18$

   i.   6 x 5 = 5 x 6 = _____

   ii.   56 x 1 = _____

   iii.   70 x 2 x 0 = _____

   iv.   3 (4 + 3) = (3 x ◯) + (3 x ◯) = _____

3.3 Parts of the Multiplication Number Sentence

The three main parts of a multiplication problem are:

❖ **Multiplicand** – the number you multiply

❖ **Multiplier** – the number by which you multiply

❖ **Product** – the result or answer

For example

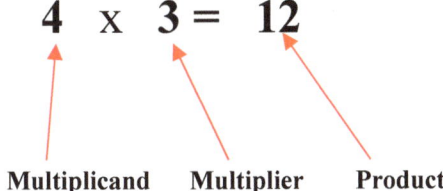

4 x **3** = **12**

Multiplicand   Multiplier   Product

For example

1 2 3 ← Multiplicand

X 2 ← Multiplier

2 4 6 ← Product

28

3.4 Multiplication of Numbers without Regrouping

When multiplying numbers, you must:

- ➢ Write the numbers down vertically if they are too big to multiply horizontally

- ➢ Ensure that each number's digit is in its correct place value column

- ➢ Then multiply the numbers, starting from the **Ones** Columns and move leftward to sequential columns. (If the sum of the numbers is less than ten, then there is no need for regrouping).

- ➢ Add all the digits in the Ones' column then put their product in the Ones' column answer space;

- ➢ Repeat the same steps for number in the Tens' column and so forth. For example:

Question: What is 2031 multiply by 2?

Th. H. T. Ones
2   0   3   1
         x 2
_____
4   0   6   2

One column: 2 x 1 = 2
Tens column: 2 x 3 = 6
Hundreds column: 2 x 0 = 0
Thousands column: 2 x 2 = 4

i). 3 4 1
　　x 2
　　_____

ii). 2 0 2 1
　　　x 3
　　　_____

iii). 1 2 2 0
　　　x 4
　　　_____

iv. 2 1 3 0
　　　x 2
　　　_____

v). 1 0 2 2
　　　x 3
　　　_____

vi). 4 8 9 5 2
　　　　x 1
　　　　_____

3.5 Multiplying Numbers with Regrouping

The system that we use is the 'Base-10 System.' Hence, ten groups of Ones equal one Ten; ten groups of Tens equal One Hundred; ten groups of a hundred equal One Thousand etc. Therefore, the largest number digit that is placed in any place value column is nine.

- When multiplying numbers, follow the same steps as stated above

- However, if the **product** of number added in any column is more than nine (9), you have to put it outside and regroup it. Your knowledge of multiplication tables will come in handy when solving these problems. For example:

Question: What is the product of 2346 and 2?

Th. H. T. Ones

+ 1
2 3 4 6   ~~12~~
   x 2
4 6 9 2

➤ In the Ones column, 2 x 6 = 12, twelve is more than nine so it is written outside and regrouped;

➤ In the Tens column, (2 x 4) + **1** = 9;

➤ The '+ **1**'is the one that was carried to the Tens column from the 12 Ones that were regrouped in the Ones column (**12 Ones** = **1** tens + 2 ones);

➤ When a number is carried to another column because of regrouping, this number is **added** to the product of the column in which it is placed.

TT. Th. H. T. Ones

+2 +1
6 0 7 5
            x 3
1 8 2 **2** **5**

❖ **Ones Column**: 3 x 5 = 15
   15 Ones = **1** ten + **5** ones
❖ **Tens Column**: (3 x 7) +1 = 22
   22Tens = **2** hundreds + **2** tens
❖ **Hundreds Column:** (3 x 0) + **2** = 2
❖ **Thousands Column:** 3 x 6 = 18
   18 Thousands = 1 ten thousand + 2 thousand

```
      Th. H. T. Ones              Th. H.  T. Ones             Th. H.  T. Ones
i).   3  1  5 9        ii).  4  6  2  1         iii).  1  2  0  3
         x  2                       x  4                      x  8
      _____                 _____                 _____

      _____                _____                 _____
```

```
      Th. H . T. Ones             Th. H.  T.  Ones          TT. Th.  H.  T.  Ones
iv.   2  1  3  0        v).  1  0  2  2          vi).  2  8  9  5 2
         x  5                     x  7                        x  3
      _____                 _____                 _____

      _____                _____                 _____
```

3.6 Multiplying Numbers by Multiples of 10s

Numbers that have two or more digits and end with zero(es) are multiples of tens. Simply put, think counting in tens or multiplying by tens. (e.g. 10, 20, 30, 100, 120, 500, 1000…)

```
Th. H. T. Ones
  3  2 4  1
     x 2 0
  _____
  6  4 8 2 0
```

```
Th. H. T. Ones
  3  2 4  1
     x 2 000
  _____
  6  4 8 2 000
```

Hint:

➤ Any number multiply by '0' equals '0'

➤ Any digit multiplied by a **multiplier** '0' will equal '0'

➤ So, bring the zero(s) down as a place holder(s)

➤ Multiply each digits of the **multiplicand** (i.e. the top number) by the multiplier that is not a '0'.

Multiplying a Number Quickly By a Multiple of Ten

$$1 \times 10 = 10$$

$$10 \times 10 = 100$$

$$10 \times 10 \times 10 = 1000$$

$$10 \times 10 \times 10 \times 10 = 10,000$$

$$20 \times 200 = 4,000$$

$$2,000 \times 300 = 600,000$$

$$30 \times 5,000 = 150,000$$

Hint: $1 \times 1 = 1$ $1 \times 1 \times 1 = 1$
$0 \times 0 = 0$ $2 \times 2 = 4$
$2 \times 3 = 6$ $3 \times 5 = 15$

➢ Count the amount of zeroes at the end in each problem and write back the same amount of zeroes at the end of your answer.

➢ For example, the problem 20 x 200 has three zeroes at the end, write them back at the end in your answer;

➢ Then multiply the digits in front of the zeroes (e.g. 2 x 2) and put the answer in front of the zeroes; 20 x 200 = 4,000

Solve these problems below correctly

1) $4 \times 200 =$

2) $600 \times 300 =$

3) $12 \times 5000 =$

4) $70 \times 80 =$

5) 5 0 1 3
       x 1 0
    _____

    _____

6) 7 3 2 0
      x 3 0 0
   _____

   _____

7) 9 1 6 5
       x 2 0
    _____

    _____

8) 1 3 8 2
      x 4 0 0 0
   _____

   _____

3.7 Long Multiplication

❖ When the **multiplier** has two or more digits with neither of them being a zero (0):

❖ Multiply the **multiplicand** first by the multiplier's digit in the 'Ones' column.

❖ Then multiply the multiplicand by the digit in the 'Tens' Column (Remember to put a '**0**' down as a placeholder to take up the 'ones' line space).

❖ Then add the two results together. See example below:

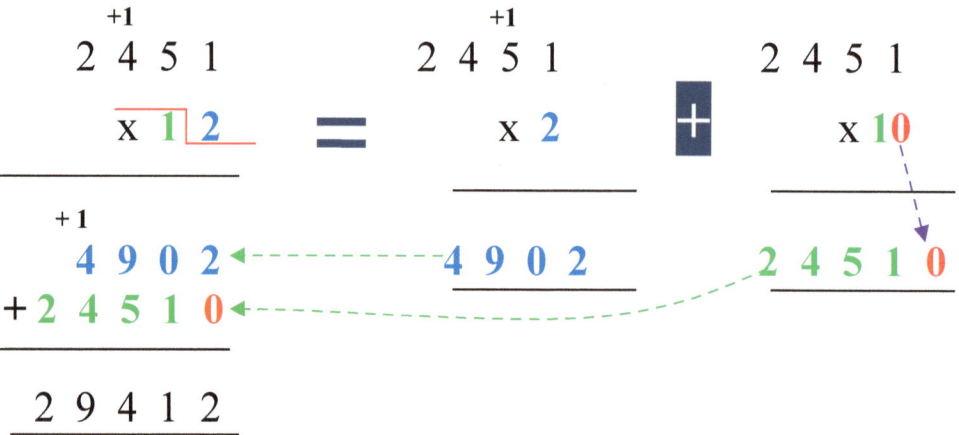

❖ If the multiplier has a digit in the Hundreds Column, you have to put two zeroes down as placeholders to take up the Ones Column and Tens Column spaces; this is so because if you are multiplying by the hundreds column then your answer should start from the hundreds place value. For example:

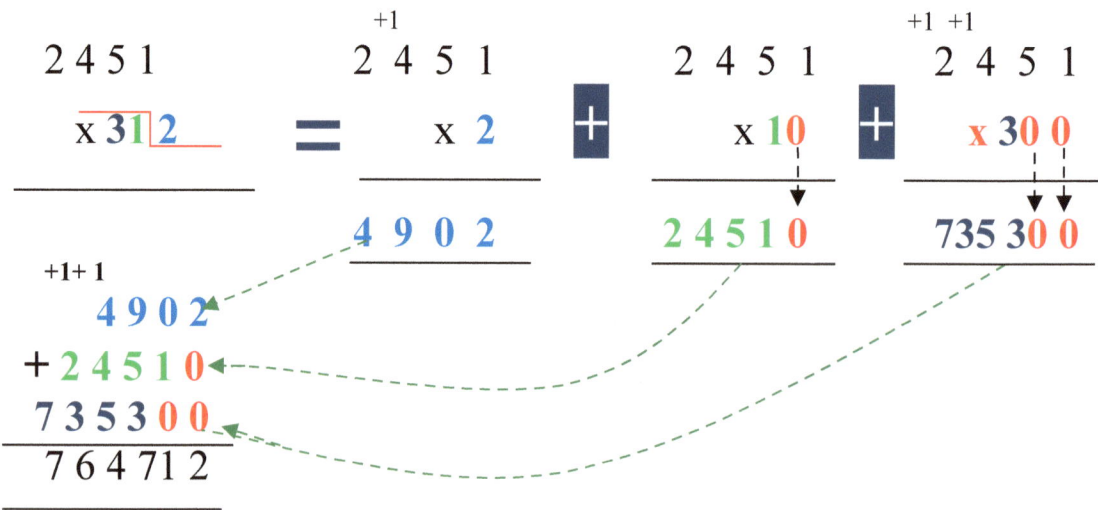

Solve these multiplication problems correctly. Show all workings.

1) 4 6 1
    x 1 1
   _____

2) 8 3 0 2
     x 6 2
   _____

3) 9 7 4 0
     x 1 4 3
   _____

4) 1 2 8 5
     x 6 0 3
   _____

5) 5 0 7 3
     x 7 6
   _____

6) 8 6 5 4 3
      x 2 4 1
   _____

Chapter 4 DIVISION

4.1 Understanding Division

Division is repeated subtraction. The division sign can be shown in several different ways. It can be shown as ÷ or $\overline{)}$ To divide means to share something equally. These concepts will be discussed in more detail below.

4.2 Properties of Division

➢ When a number is divided by itself, the result is one (1).
   For example: $12 \div 12 = 1$ $76 \div 76 = 1$

$$12\overline{)12}^{\,1} \qquad\qquad 76\overline{)76}^{\,1}$$

➢ Any number that is divided by one (1) is equal to itself.
   For example $25 \div 1 = 25$ $300 \div 1 = 300$

$$1\overline{)25}^{\,25} \qquad\qquad 1\overline{)300}^{\,300}$$

4.3 Parts of the Division Number Sentence

The three main parts of a division number sentence are the _Dividend_, the _Divisor_ and the _Quotient_.

Dividend: This is the number that is being divided (the 'Have' number)

Divisor: This is the number (e.g. the number of persons) that the dividend is being divided or shared with

Quotient: This is the number that tells how much each person got

For example

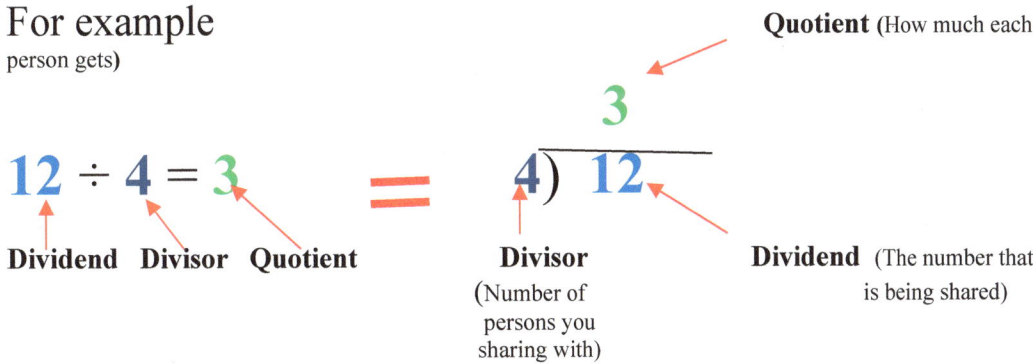

N.B.
Please note the position of each part of the division number sentence in the methods used above and their simplified meanings.

4.4 Relationship Between Division and Multiplication

Division is the opposite of multiplication. Study the tables below that show the relationship between multiplication and division.

Multiplication		Division
$2 \times 1 = 2$	2 into 2 goes 1	$2 \overline{)2}$ quotient 1
$2 \times 2 = 4$	2 into 4 goes 2	$2 \overline{)4}$ quotient 2
$2 \times 3 = 6$	2 into 6 goes 3	$2 \overline{)6}$ quotient 3
$2 \times 4 = 8$	2 into 8 goes 4	$2 \overline{)8}$ quotient 4
$2 \times 5 = 10$	2 into 10 goes 5	$2 \overline{)10}$ quotient 5
$2 \times 11 = 22$	2 into 22 goes 11	$2 \overline{)22}$ quotient 11
$2 \times 12 = 24$	2 into 24 goes 12	$2 \overline{)24}$ quotient 12

4.5 Finding Any Missing Part of a Division Number Sentence

$$\boxed{?} \div \boxed{?} = \boxed{?}$$

Dividend   **Divisor**   **Quotient**

➢ To find the **Quotient**, divide the Dividend (Have Number) by the Divisor (number of people you are sharing with)   e.g. $15 \div 3 = \boxed{?}$

$$3)\overline{15}\quad \boxed{5}$$

➢ To find the **Divisor,** divide the Dividend by the Quotient   e.g. $21 \div \boxed{?} = 3$

$$3)\overline{21}\quad \boxed{7}$$

➢ To find the **Dividend,** multiply the Divisor by the Quotient   e.g. $\boxed{} \div 4 = 5$

$$\boxed{} = 5 \times 4 = 20$$

State the part of each division number sentence which is missing. Then carry operation to solve it.

1) $\boxed{} \div 8 = 3$ 2) $42 \div 6 = \boxed{}$ 3) $66 \div \boxed{} = 11$

4) $26 \div \boxed{} = 2$ 5) $\boxed{} \div 3 = 8$ 6) $81 \div 9 = \boxed{}$

4.6 Dividing Numbers Without Regrouping

➢ When dividing a number without regrouping, it means that the digit(s) of the number (dividend) can be divided into a given number (divisor) without leaving a remainder.

➢ When dividing a number (i.e. dividend), you should start with the digit that is in the highest place value position.

➢ Then continue dividing each of its digits, working from left to right until you reach the last digit. For example

$$2\overline{)462} \quad \begin{array}{c} 2\,3\,1 \end{array}$$

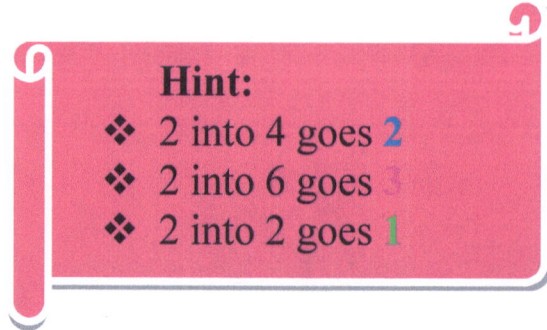

Hint:
❖ 2 into 4 goes **2**
❖ 2 into 6 goes **3**
❖ 2 into 2 goes **1**

1. $3)\overline{633}$ 2. $4)\overline{448}$ 3. $2)\overline{8262}$

4. $5)\overline{5555}$ 5. $3)\overline{396}$ 6. $2)\overline{4426}$

4.7 Dividing Numbers with Regrouping

When dividing a number with regrouping, it means that one or more digits of the number (dividend) cannot be divided by another number (divisor) without a remainder.

➢ As mentioned before, begin dividing the dividend (Have number) by starting with the digit in the highest place value position.

➢ If there is a remainder, add it to the next digit

➢ Continue dividing each digit of the dividend, including the change in value of any of its digits.

➢ If there is a remainder when you divide the last digit, write it at the end of the quotient as **'r 1'** ; this is short for 'remainder 1'. For example

What is 357 divided by 2?

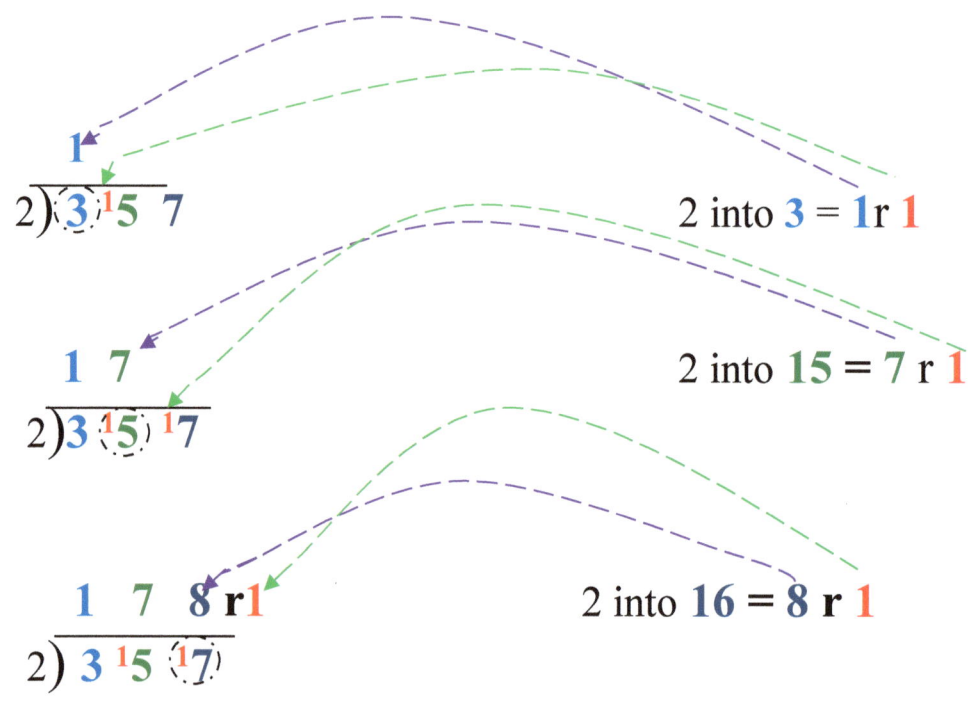

2 into **3** = **1**r **1**

2 into **15** = **7** r **1**

2 into **16** = **8** r **1**

Solve each problem below correctly

1. 3)741

2. 4)458

3. 2)1383

4. 6)7388

5. 3)2396

6. 2)3021

4.8 Long Division

The division problems above can be referred to as *'short division'*. In long division, the same thought processes take place. However, it goes a step further by showing the total value of each part of dividend that is given away or shared vertically (downward) and how much remains. For example

- ❖ The **quotient** (how much each person got) is shown at the top. This is solved similar to the steps in the short division above.

$$\begin{array}{r} 1 \\ 3\overline{)4\ 1} \end{array}$$

- ❖ The **value** of the part of the dividend that is divided, is calculated easily by multiplying the *divisor* by the digit placed as part of the *quotient.* This multiplication is shown by the side. The result from this multiplication is then subtracted from the specified digit of the dividend.

$$\begin{array}{r} 1 \\ 3\overline{)4\ 1} \\ (3 \times 1) \quad -\underline{3} \end{array}$$ 3 into 4 = 1r1

- ❖ The remainder, if any, is written below the subtraction problem

$$\begin{array}{r} 1 \\ 3\overline{)4\ 1} \\ (3 \times 1) \quad -\underline{3} \\ 1 \end{array}$$ 3 into 4 = 1r1

➢ The next digit of the dividend is carried downward. Together with the remainder, this is treated as the next dividend part to divide.

```
              1 3
          3)4 1          ← — — —   3 into 4 = 1r1
(3 x 1)    -3
           11            ← — — —   3 into 11 = 3 r 2
(3 x 3)   - 9
            2
```

➢ The remainder, if any, is again written at the bottom and the steps are repeated until the last digit of the dividend is divided.

Exercise 17

Solve EACH of these problems using both the short division and long division methods:

1. 2)53

4. 12)382

2. 4)49
3. 9)190

4.9 Solving Worded Problems

When solving worded problems, there are certain key words that will give you hint on what mathematical operation is required. Some key words that mean to divide are listed below.

Words That Mean To Divide

- ❖ share (equally)
- ❖ quotient
- ❖ How many groups of ..are found in ....?

- ❖ How many did each get?
- ❖ What is the divisor of..
- ❖ Divide ... by

Sometimes, a worded problem that requires you to divide may not contain any of these phrase listed above. However, you must remember that division is **'repeated subtraction,'** that is, the same number is taken away or placed in a group each time. For example

Sixty-five pies were **shared** equally among **seven persons**. How many pies did each person receive?

$$\begin{array}{r} 9 \text{ R } 2 \\ 7\overline{)65} \end{array} \qquad \text{Answer} = 9\tfrac{2}{7}$$

Each person received **9** whole pies and **2/7** of a pie.

(N.B. the remainder is written as a fraction of the divisor)

Solve EACH of these problems correctly

1. A shopkeeper had 306 trays of eggs. He sold them to some buyers. If each buyer received 6 trays of eggs, how many buyers did he sell the eggs to?

2. What is the quotient of 9854 and 6?

3. A school consists of 7000 students. If 35 students are placed in each class, how many classes does the school have?

4. Divide $28,434 equally among 7 persons.

5.1   Solving Problems Involving More Than One Mathematical Operations:

Sometimes you may be given problems that require you to use more than one Mathematical operation to solve them. Some common rules to follow are represented by the acronym 'BODMAS' which stands for:

➤ Do all problems in Brackets first;

➤ then 'Of',

➤ then Division and Multiplication;

➤ then afterwards, solve the Addition and Subtraction.

For example: Solve: $12 + 3 - 4 \times 2 + (9 + 6) \div 5$

✓ **Rule 1**: *Do bracket first*
$12 + 3 - 4 \times 2 + (9 + 6) \div 5$
$= 12 + 3 - 4 \times 2 + 15 \div 5$

✓ **Rule 2:** *Solve division and/or multiplication next*
$12 + 3 - 4 \times 2 + 15 \div 5$
$= 12 + 3 - 8 + 3$

✓ **Rule 3:** *Finally, solve addition and subtraction*

$12 + 3 - 8 + 3$
$= 5 - 8 + 3$
$= 7 + 3 = 10$

1). $34 + (81 - 63) =$

2). $14 + 6 \times 3 =$

3). $72 + 12 \div 4 \times 5 - 9 =$

4). $1000 - 600 + (18 - 5) =$

5). $1764 + 459 \div 9 + 2 \times 5 =$

6). $3 \times 5 \times 2 - (4 + 5) =$

THINGS YOU NEED TO REMEMBER

➢ The four (4) mathematical operations are: **Addition, Subtraction, Multiplication** and **Division.**

Addition

➢ Some words that mean to add are: *sum, plus, add, increase, total, got more, altogether* and *in all.*

Properties of Addition:

➢ **Identity Property:** If you add zero (0) to any number, it will remain the same. Example: 5 + 0 = 5, 790 + 0 = 790

➢ **Commutative Property**: It does not matter which addend is added first, what is important is that the numeral digits are placed in their correct place value column before adding.

➢ **Associative Property**

The associative property shows that the sum of addends (the answer) remains the same regardless of how they are grouped.

➢ **Adding Numbers:** When adding numbers remember It does not matter which number is added first. What is important is that each number's digit is in its correct place value column.

➤ If the numbers of a column add up to be more than '9,' put the answer outside and regroup it.

Adding Numbers with Decimals: When adding numbers with decimal remember

➤ The decimal point separates the whole number from the fraction/part of a whole.

➤ All the digits are placed in their correct place value column.

➤ All the decimal points are placed directly under each other.

➤ If some numbers have two digits after the decimal point and others do not, you can put a '0' in their tenths or hundredths column to make adding the numbers easier.

Finding the missing part of an addition number sentence:

➤ The main numbers in an addition problem are the **addends** and the **sum.**

➤ The **sum** is the **largest number** in an addition problem. It usually comes at the end of the problem.

$$(e.g.\ 14 + 35 = 49)$$

addends   sum

➢ To find the sum, **add the addends** together.

$$(e.g. \ 7 + 8 = \underline{\quad})$$

➢ To find the missing addend, **subtract the given addend**(s) from the sum.

➢ If the numbers in the problem are too big to work out horizontally, write them down vertically and solve correctly.

Subtraction

➢ Some words that mean to subtract are: *minus, difference between, less than, more than ___ than (comparing), decrease, reduce, take ___ from, left and remaining.*

➢ Always write the biggest number first.

➢ Write each number's digit in its correct place value column.

➢ Only borrow (take from the next column) if the digit to be taken away is **bigger** than the 'have' number in that column. For example:

What is the difference between 6583 and 899?

Th. H. T. Ones

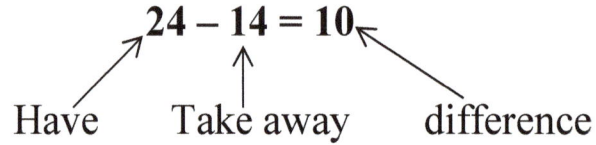

```
      5  14
   6  5 ¹8  3 ---- Have number
 -     8  9  3 ---- Take away number
   5  6  9  0 ---- Difference or remainder
```

24 – 14 = 10

Have   Take away   difference

> The **largest number** always comes at the beginning (at the top). This can be seen as the number of things you (or the persons) have.

> The middle number is the amount that is being taken away.

> The **difference** or **remainder** is the answer showing what remains after subtraction. This number is smaller than the 'Have' number and comes at the end.

> The opposite of subtraction is addition.

To find the 'Have' number, add the number taken away and the remainder. Example:

$$\text{If } W - 4 = 5, \text{ what is } W? \qquad W = 5 + 4 = 9$$

> ➤ To find the 'difference, subtract the 'taken away' number from the 'Have' number. Example:

$$\text{If } 22 - 20 = n, \text{ what is } n? \qquad n = 2$$

> ➤ To find the 'take away' number, subtract the difference from the 'Have' number. Example:

$$\text{If } 9 - b = 7, \text{ what is } b? \qquad b = 9 - 7 =$$

Multiplications

Multiplication is repeated addition; that is you are adding the same number each time.

Some words that mean to multiply are: *product, times, multiples, 6each...together, each...in all...*

Properties of multiplication are:

➤ **Commutative**   e.g. 2 x 4 = 4 x 2

➤ **Associative**   e.g. 2 x (4 x 3) = (2 x 4) x 3

➤ **Identity** - any number multiply by '1' equal itself
  e.g. 6 x 1 = 6

➤ **Multiplying by '0'** – any number multiply by '0' equal '0'
  e.g. 2 x 4 x 0 = 0

➤ **Distributive:** multiplying a number by the sum of two or more numbers is the same as multiplying each addend by the number ten adding their products.

The three main parts of a multiplication problem are:

❖ **Multiplicand** – the number you multiply

❖ **Multiplier** - The number by which you multiply

❖ **Product** – the result or answer

```
    1 2 3 ◄——— Multiplicand
      X 2 ◄——— Multiplier
    2 4 6 ◄——— Product
```

Long Multiplication:

When the multiplier has two or more digits with neither of them being a zero (0): Multiply the multiplicand first by the multiplier's digit in the 'ones' column. Then multiply the multiplicand by the digit in the 'tens' column (Remember to put a '0' down to take up the 'ones' line space). Add the two results together. For example:

What is 2451 multiply by 12?

```
        +1
     2 4 5 1
      x 1 2
     _____
        1
     4 9 0 2
  +2 4 5 1 0
  _____

   2 9 4 1 2
  _____
```

Division

Division is repeated subtraction; that is you are taking away the same number each time.

Some words that mean to divide are:
divide, share, quotient, decimal fraction, put ---in each, give each … (repeated subtraction) …

Solving Problems with more than one Mathematical Operations

When solving problems with more than one mathematical operations use **BODMAS**' which is an acronym for: Do all problems in **B**rackets first; then **'Of'**, then **D**ivision and **M**ultiplication, then afterwards, solve the **A**ddition and **S**ubtraction

OPERATIONS AND RELATIONS

MASTERY TEST 1

NAME: _____ DATE: ___/___/_____
                                        DD   MM   YYYY

Solve these problems correctly. Show all workings:

1. Add: 2870, 146 and 109

2. Multiply 4063 by 70

3. Minus: $78.00 from $102.00

4. Divide 5251 by 3

5. Add: 34.65, 12.03 and 5.8

6. Subtract: 69.43 from 90.1

7. What is the sum of 5678 and 1040?

8. Joe spent $37.15 from $50. What was his change?

9. Marlene has 100 mangoes. Sally has 53 mangoes. How many more mangoes does Marlene have than Sally?

10. If 693 students were to be placed equally in 7 classes. Approximately how many students will be placed in each class?

11. A shopkeeper sold 906 trays of eggs. If each tray contained 20 eggs, how many eggs were sold in all?

12. What is the product of 9607 and 69?

13. What is 5429 divided by 24?

14. Sixty-five pies were shared equally among seven persons. How many pies did each person receive?

15. Solve:

A. $5 + (8 \times 7) - 35 =$

B. $63 \div 3 \times 4 - 12 =$

C. $321 - 890 \div 10 + (15 \times 3) =$

D. $75 + 9 \times 5 - (125 \div 5) =$

OPERATIONS AND RELATIONS

MASTERY TEST 2

NAME: _____   DATE: ___/___/___

Solve these problems correctly. Show all workings:

1.   223
    +130

2.   5006
    + 4731

3. 3922
    x2

4.   6745
    - 403

5.   2301
    x 13

6.   8000
    - 754

7. Multiply 9221 by 60

8. Minus: $57.04 from $100

9. Divide 4582 by 4

10. Find the difference between 8735 and 6480

11. Add: 89.65, 2.09 and 305.28

12. Subtract: 43.43 from 50.11

13. What is the sum of 4077 and 1264?

14. Joe spent $25.15 from $60. What was his change?

15. Donald has 206 pizzas. Nelly has 153 pizzas. How many more pizzas does Donald have than Nelly?

16. If 455 students were to be placed equally in 7 classes. Approximately how many students will be placed in each class?

17. One thousand five hundred and twenty-five banana pies were shared equally among five persons. How many pies did each person receive?

18. A man sold 8380 trays of eggs. If each tray contained 20 eggs, how many eggs were sold in all?

19. What is the product of 7802 and 36?

20. What is 4422 divided by 22?

21. Solve: $89 + 65 \times 3 + 25 \div 5 - (145 - 63) =$

NUMERACY TEST

NAME: _____ **DATE**: ___ / ___ / _____
DD   MM   YYYY

SECTION 1: NUMBER CONCEPTS

Read all questions carefully, then solve them correctly

1. What is the place value of the digit 4 in the number 7046?

_____

2. Write the place value of the digit 3 in each number below:

a. 1963   b. 3005   c. 45.30   d.
246.23

3. In the number 4678.2, what is the **value** of the digits:

a. 2   b. 4   c. 6   d.7
e. 8

4. Arrange these numbers below in order of sizes from smallest to largest

a. 245,   225,   451,   199,   189

_____

b. 1.3,   1.02,   1.45,   1.29,   1.00,   1.90

_____

c. ½,   1/3,   1/9,   ¼,   1/10 , 1

_____

d. 3/8, 1/8, 7/8, 4/8, 6/8, 0/8

_____

5. Compare these number using the signs: <   =   >

(a). 358   359   (b). 3/6   8/18   (c). 2.19   2.3

(d). ix   xi   (e). 2/2   3/3   (f). $50.50 $505.0

(g). 2/10   0.2   (h). 10 x10   20+20   (i). 450c $4.25

6. Write the names for these numbers:

a. 5096

_____

b. 2012

_____

c. XLII

_____

d. 2.56

_____

e. 12th

_____

f. 40th

_____

7. Write these numbers in figures:

a. eight thousand and one _____

b. one hundred and twentieth _____

c. thirty-five thousand _____

d. seventeen point six seven _____

8. List the factors of 15   { _____ }

9. What is the L.C.M. of 4, 6 and 24

10. Expand this number: 67095

_____

11. Complete these number sequences:

a. 98, 97, ____, ____   b. 2, 5, ____, ____   c. ____, ____, 21, 28

d. 1.5, 1.6, ____, ____   e. iii, vi, ____, ____   f. ____, 4th, 6th,

____

SECTION 2: **OPERATIONS AND RELATIONS**

Solve these problems correctly. Show all workings:

1. Add: 2870, 146 and 109

2. Multiply 4063 by 70

3. Minus: $78.00 from $ 102.00

4. Divide 5263 by 16

5. Increase: 45.65 and 12.03 by 5.8

6. Subtract: 69.43 from 90

7. From the sum of 3678 and 1032, take their differences

8. Joe spent $37.15 from $50. What was his change?

8. Marlene has 100 oranges. Sally has 53 oranges. How many more oranges does Marlene have than Sally ?

9. If 1694 students were to be placed equally in 8 classes. Approximately how many students will be placed in each class?

10. Sixty-five cup-cakes were shared equally among seven persons. How many cup-cakes did each person receive?

NUMERACY TEST

NAME: _____ DATE:___/___/_____
DD MM YYYY

SECTION 1: **NUMBER CONCEPTS**

Read all questions carefully, then solve them correctly

1. What is the place value of the digit 4 in the number 7046?

_____

2. Write the place value of the digit 3 in each number below:

a. 1963   b. 3005   c. 45.30   d. 246.23

3. In the number 4678.2, what is the **value** of the digits:

a. 2   b. 4   c. 6   d.7   e. 8

4. Arrange these numbers below in order of sizes from smallest to largest

a. 245, 225, 451, 199, 189

_____

b. 1.3, 1.02, 1.45, 1.29, 1.00, 1.90

_____

c. ½, 1/3, 1/9, ¼, 1/10 , 1

_____

d. 3/8, 1/8, 7/8, 4/8, 6/8, 0/8

_____

5. Compare these number using the signs: <   =   >

(a). 358   359   (b). 3/6   8/18   (c). 2.19   2.3

(d). ix   xi   (e). 2/2   3/3   (f). $50.50 $505.0

(g). 2/10   0.2   (h). 10 x10   20+20   (i). 450c $4.25

6. Write the names for these numbers:

a. 5096

_____

b. 2012

_____

c. XLII

_____

d. 2.56

_____

e. 12th

_____

f. 40th

_____

7. Write these numbers in figures:

a. eight thousand and one _____

b. one hundred and twentieth _____

c. thirty-five thousand _____

d. seventeen point six seven _____

8. List the factors of 15   { _____ }

9. What is the L.C.M. of 4, 6 and 24

10. Expand this number: 67095

_____

11. Complete these number sequences:

a. 98, 97, ___, ___   b. 2, 5, ___,___   c. ___,____, 21, 28

d. 1.5, 1.6, ___,___   e. iii, vi, ___, ___   f. ___, 4th, 6th, ___

SECTION 2: OPERATIONS AND RELATIONS

Solve these problems correctly. Show all workings:

1. Add: 2870, 146 and 109

2. Multiply 4063 by 70

3. Minus: $78.00 from $ 102.00

4. Divide 5263 by 16

5. Increase: 45.65 and 12.03 by 5.8

6. Subtract: 69.43 from 90

7. From the sum of 3678 and 1032, take their differences

8. Joe spent $37.15 from $50. What was   his change?

9. Marlene has 100 mangoes. Sally has 53 mangoes. How many more mangoes does Marlene have than sally?

10. Sixty-five pies were shared equally among seven persons. How many pies did each person receive?

11. A shopkeeper sold 686 bags of rice. If each bag contained 46 pounds, how many pounds were sold in all?

CERTIFICATE OF ACHIEVEMENT

in

Mathematical Operations and Relations

at Grade _____ level

This certificate is presented to

_____

on _____ day of _____, 20_____

Signed by   Signed by

_____   _____

INDEX

NOTES

www.ingramcontent.com/pod-product-compliance
Lightning Source LLC
Chambersburg PA
CBHW050740180526
45159CB00003B/1292